HEAT RULES

beginning Thermodynamics for kids and their parents that slept through science

Gayle Weyher

Other Visual Learning Books
by Gayle Weyher

WHAT'S THAT?
Toddler Science

Beautiful Math
visual study of shapes, quantities, and patterns

Symbols

ISBN: 9798879709780

for

Elliot, Evelyn, Asher, and Allister

Heat Rules

Is life possible without heat? NO. You eat to stay alive.

Life means you metabolize and reproduce - put simply you hEAT and have rabbits. That's as simple as it gets. All organisms require heat for survival. It sustains life, and its absence is tantamount to death.

We know that HEAT is motion, ceaseless agitation of atoms and molecules that compose all matter. Your body heats as you convert the energy from the food you eat to energy that your brain and body need to function.
You do not want to sleep through science class!

Ideas and information about Thermodynamics are constantly evolving. This book is about how things are perceived now.
It is not a complete story but a solid beginning.
The illustrations included are imaginal representations
to help visualize and remember the concepts.

What is Thermodynamics?

Thermodynamics describes
energy and its interactions with matter.

Thermodynamics is a branch of physics that deals with
heat, work, and temperature
and their relation to
energy, entropy, radiation,
and the physical properties of matter -
temperature, pressure, volume, composition and time.

Thermodynamic Systems are defined as open or closed.
Open systems interact with their surroundings
feeding on the flux of matter and energy
coming to them from outside.
Closed systems are fixed by boundary conditions.

Thermodynamic Systems

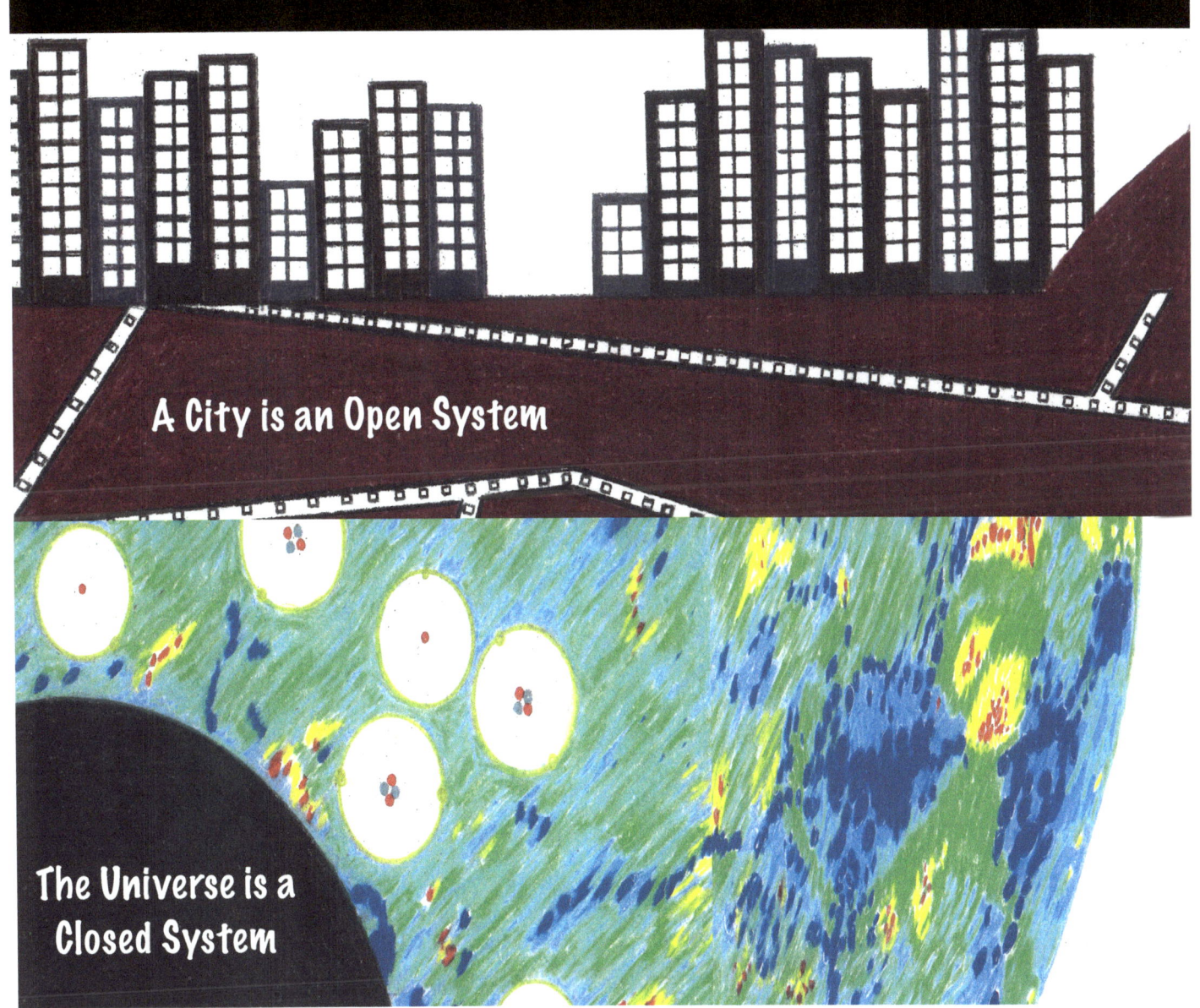

A City is an Open System

The Universe is a Closed System

What is Energy?

Energy is defined as the ability to cause something to move.
When you eat your molecules move, you gain energy,
and you heat up.

Energy is a system's potential to do work.
You need energy to do your schoolwork.

Energy is around you in different forms

kinetic energy
thermal energy
radiant energy
electrical energy
potential energy

Energy and Matter are equivalent

$$E = mc^2$$

Einstein showed in his famous equation that
Energy and Mass are different forms
of the same fundamental entity.

Energy can change its form.

How does Energy change its form?

Science shows that nature is full of changes of matter into energy and energy into matter.

A Thermodynamic System can change in only two ways: heating or cooling, and work done on or by it.

After the Big Bang, an event released a burst of light that at the first time, could permeate the Cosmos.
The Cosmic Microwave Background (CMB) reveals the remnants of the oldest known light in the Universe.

The hot expanding Universe cooled, and the primordial energy in the form of a soup of primitive quarks and gluons eventually changed into atoms.

Cosmology

Our Solar System shows the change of matter
within the sun's interior into
energy in the form of light
that animates our world.

Astrophysics

Through fission and fusion,
matter becomes energy in atomic bombs
and nuclear power plants
by changing a tiny percentage of mass
into heat and radiation.

Nuclear Technology

Light is a form of pure, matter-less energy.

Light beam accelerators change the kinetic energy
of extremely fast-moving light particles
into masses of heavy new particles by collisions.

Particle Physics

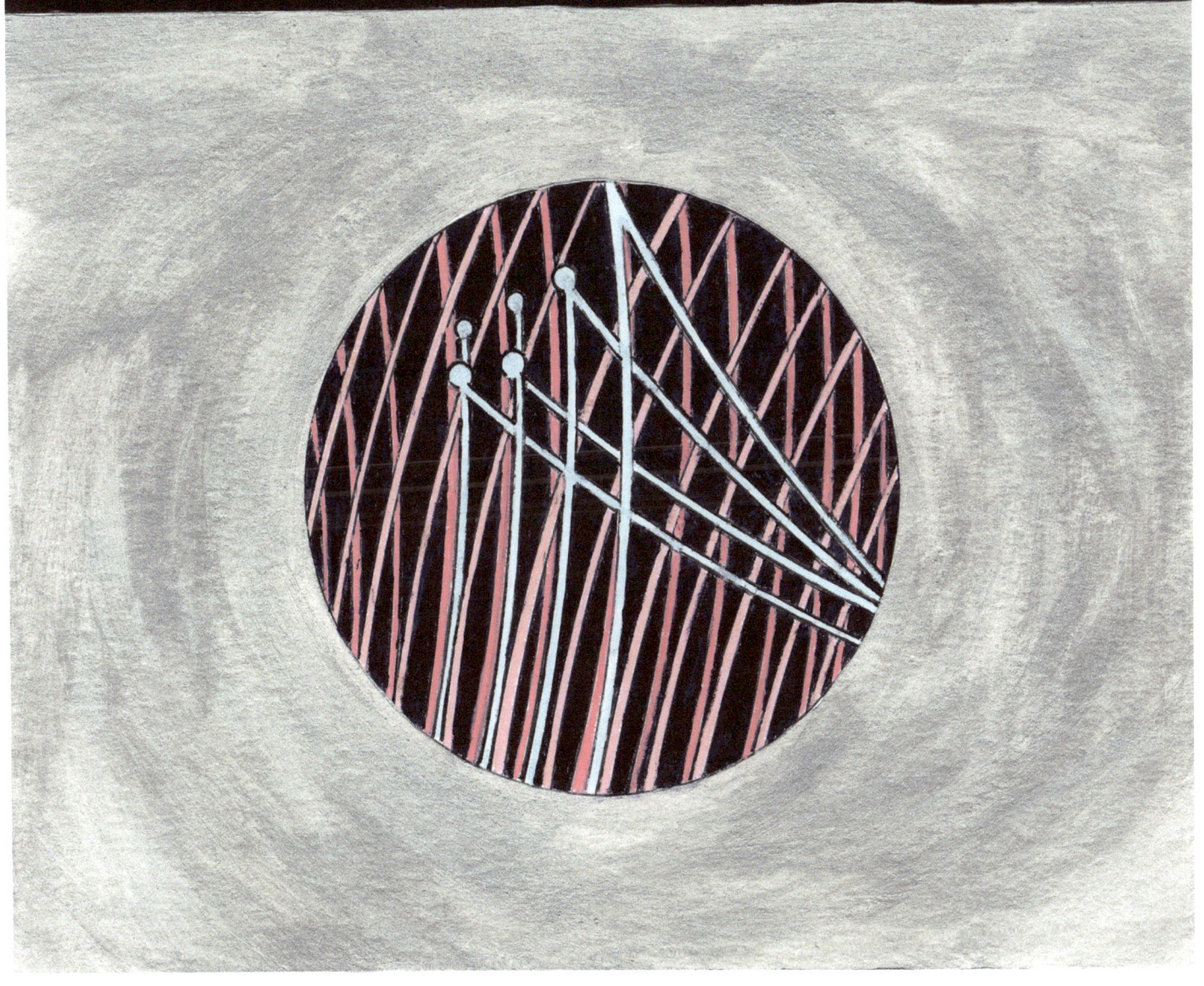

In the Earth's crust,
the change of matter into energy constantly continues
through the decay of radioactive elements.
It liberates the heat that warms us.

This is called radioactive decay. There are 28 naturally
occurring chemical elements on Earth that are radioactive.

A well-known example is Uranium.

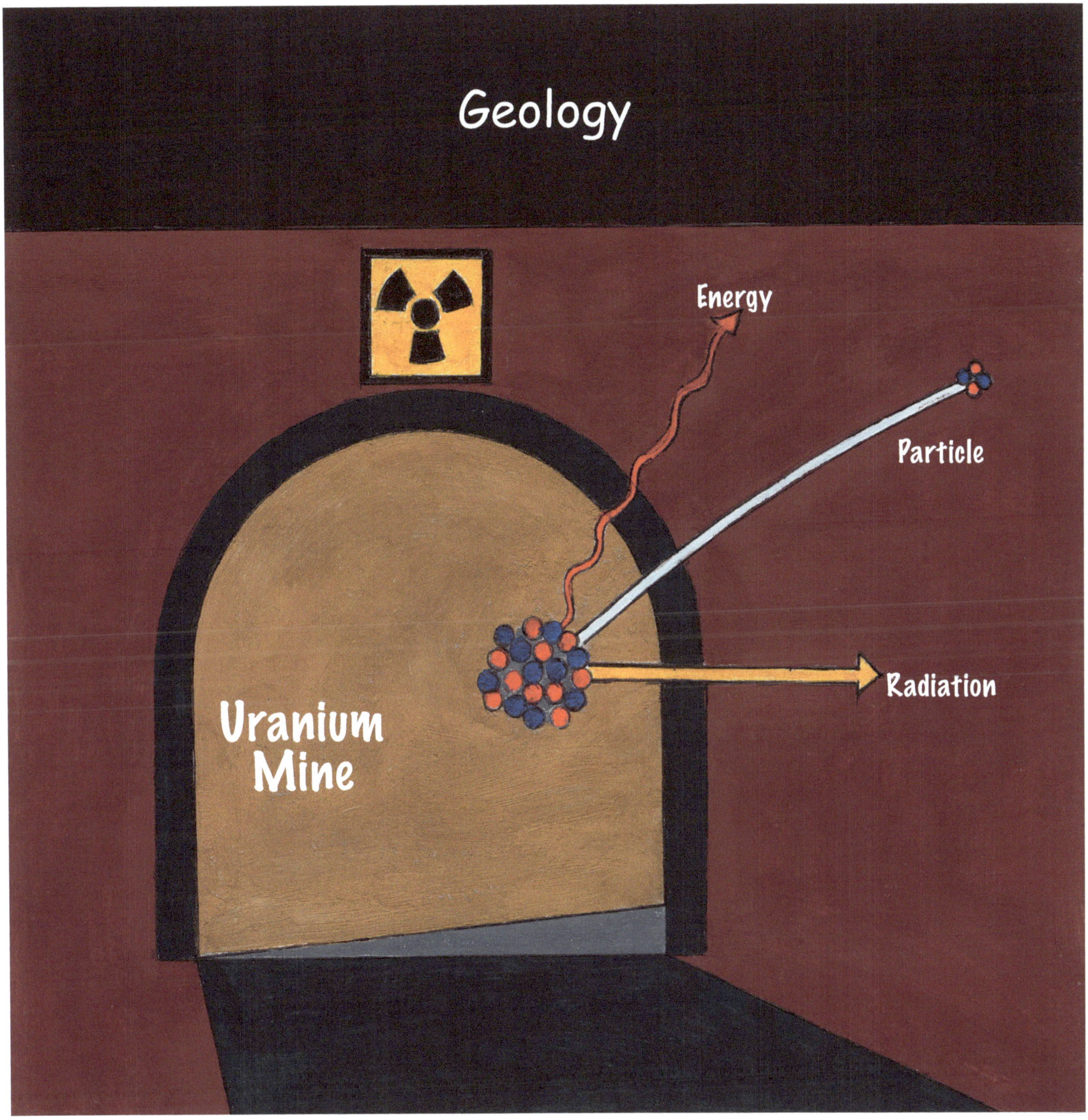

Geology

Energy

Particle

Radiation

Uranium
Mine

From a peaceful lake containing potential energy
formed by a hydroelectric dam on a river,
water flows through its intake to turbines below.

The turbines drive an electric generator
to produce electrical energy.

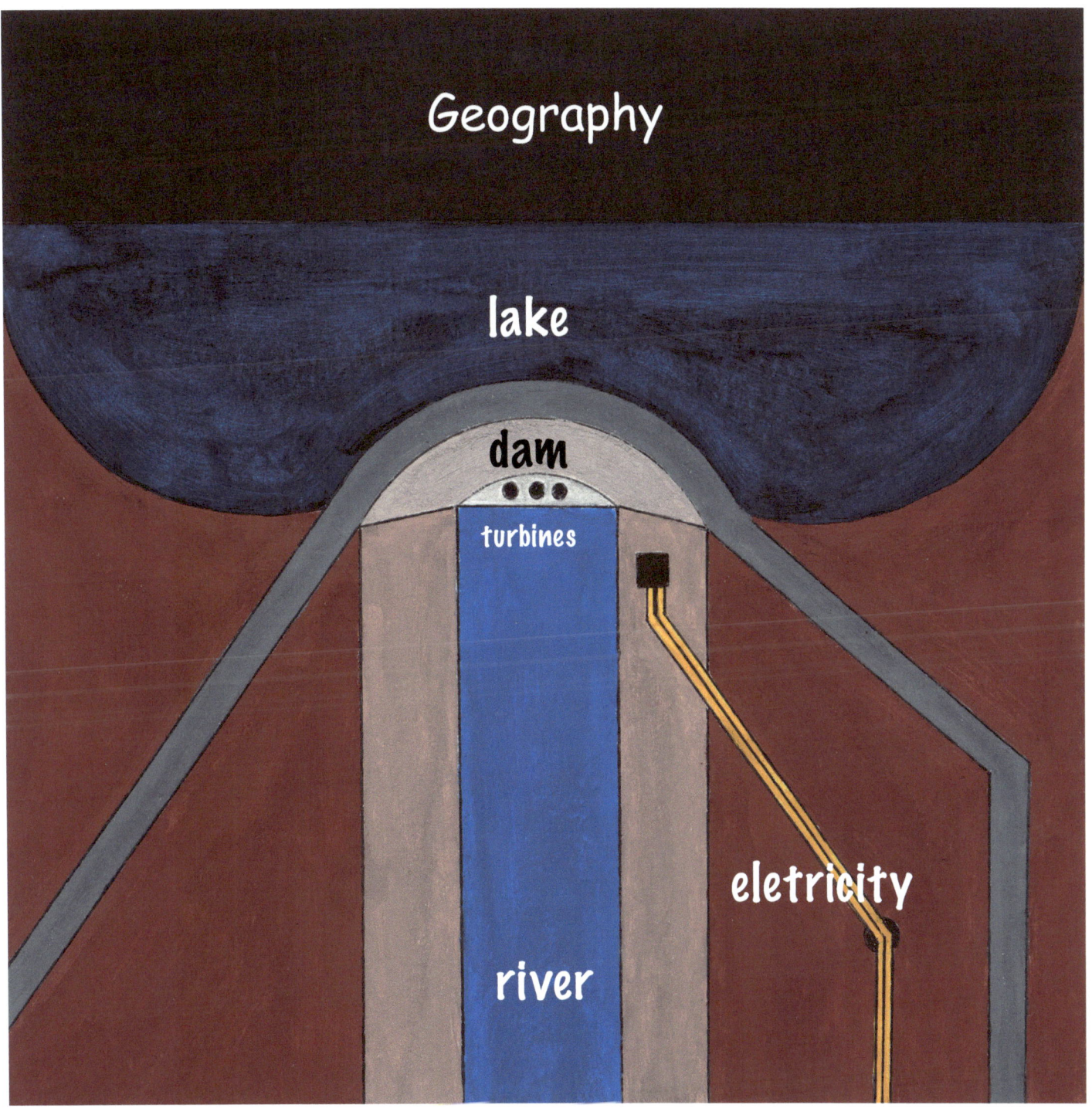

Electricity is a form of kinetic energy
caused by electrical charges.

Electrons move through metallic wires
so you can play your electric guitar
and listen to your favorite music.

Electricity

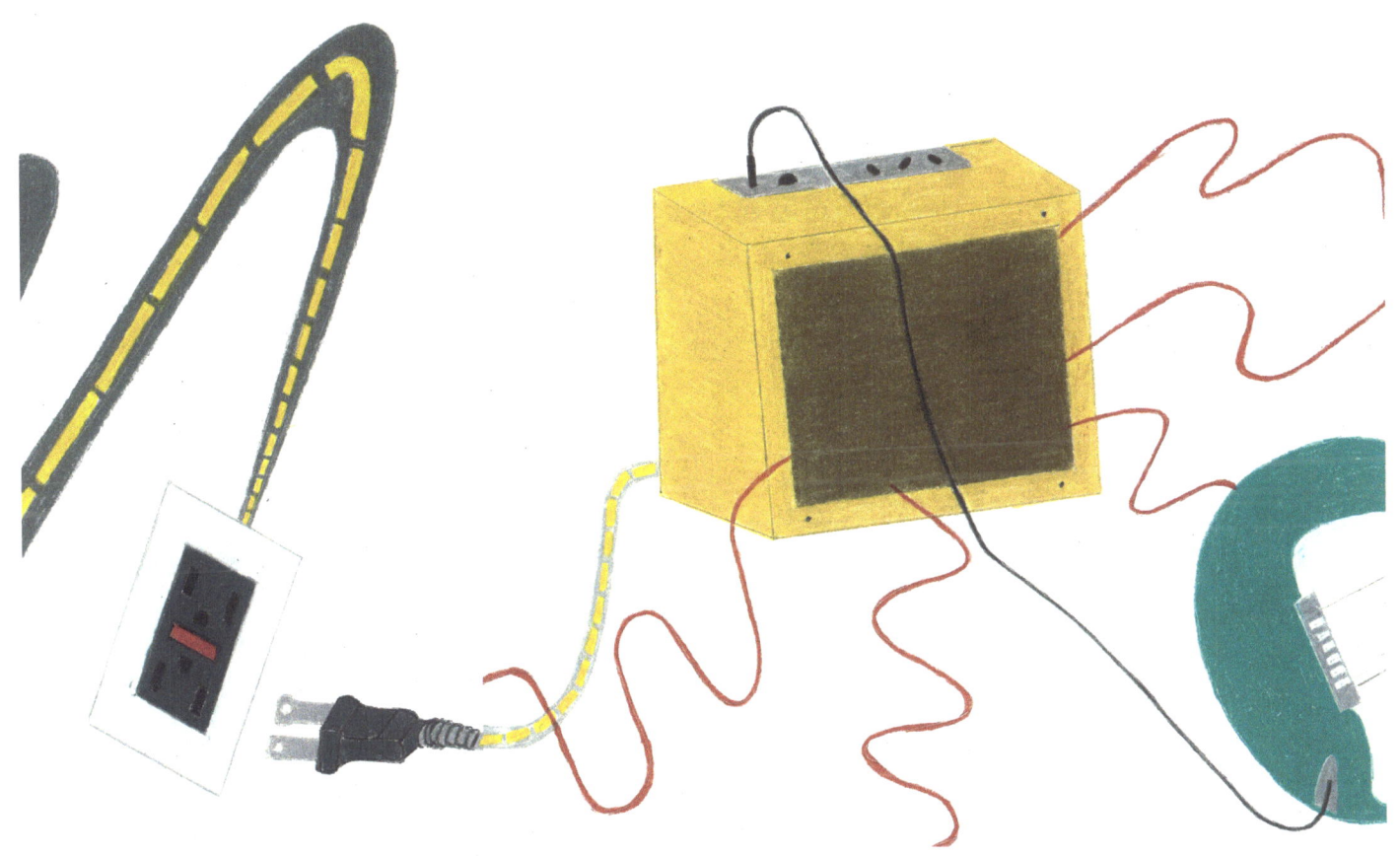

What is Heat?

Heat produces motion and motion produces heat.

Heat is the speed of the motion of particles.
The faster they move the hotter they become.

All energy unless it is stored turns to heat.

Energy is stored in gasoline which when heated
changes to work, so you can drive your car.
It also gives off heat into the environment.

It is impossible to extract work from heat
without at the same time discarding some heat.

Every time a car is driven, 2/3 of the fuel is wasted.

Heat

Heat expands the volume of any gas.

Hot air balloons contain heated air.

They float because the air inside the balloons
is a lower density
than the cooler air outside them.

Hot Air Balloons Rise

Air rises when the land heats up.

Air contains water vapor - water in gas form.

The amount of water vapor in the air
depends on the air temperature.

Warmer air holds more water vapor.

When warm, moist air rises
clouds form as the air cools.

Air is also forced upward when
it comes in contact with mountains.

Clouds Rise

When a meteor falls from space
and enters the Earth's atmosphere,
the friction of the rock
rubbing against the atmosphere
causes heat and light
like a fire.

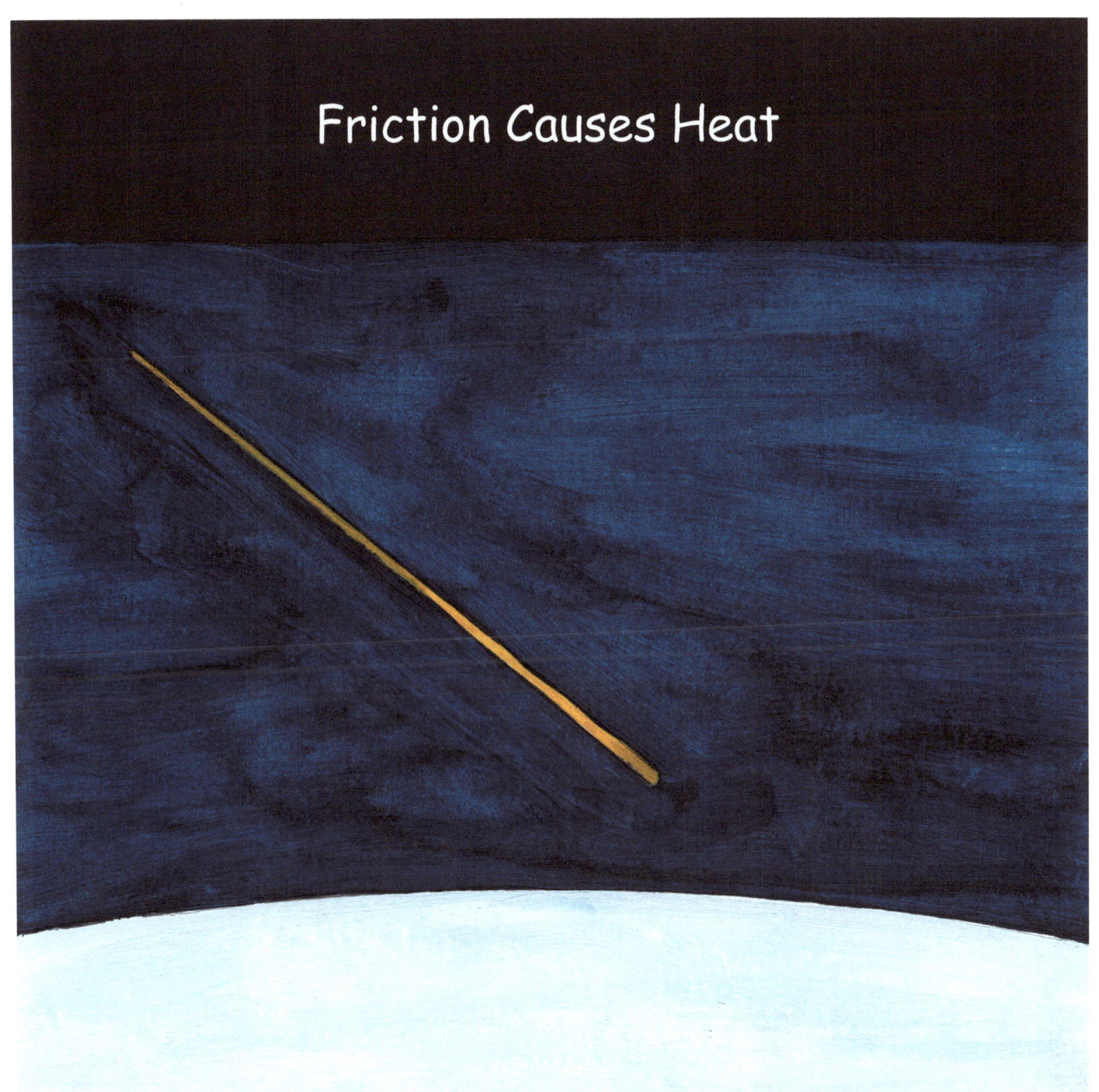

The hot cocoa molecules in the cup bump into
the colder air they come in contact with.

The hot molecules transfer their energy to the cold molecules.
Eventually the cocoa and air molecules
will be the same temperature.

Hot Cocoa Cools

The chemical energy in food is changed by your body
into moving mechanical energy and heat energy.

Chemical energy is stored energy (potential energy).
It is stored in the bonds between atoms and molecules.

Chemical energy is what holds the atoms in a molecule together.

It is also what holds together molecules in a substance
like French Fries.

When you eat you HEAT

French Fries

Thermal radiation is one of the fundamental mechanisms of heat transfer.

Thermal radiation is seen in visible light as a glow on hot metal.

Thermal radiation is electromagnetic radiation generated by the heat motion of particles in matter.

All matter with a temperature greater than absolute zero emits thermal radiation.

Thermal Radiation

On Earth, uranium and other radioactive elements release nuclear energy in the form of heat.

The heat generated deep within the planet rises
and drives the currents that circulate
through the hot solid rock of the mantle
and keep it moving very slowly.

The mantle currents rise beneath the crust,
then move sideways and sink as they cool.

Mantle Currents

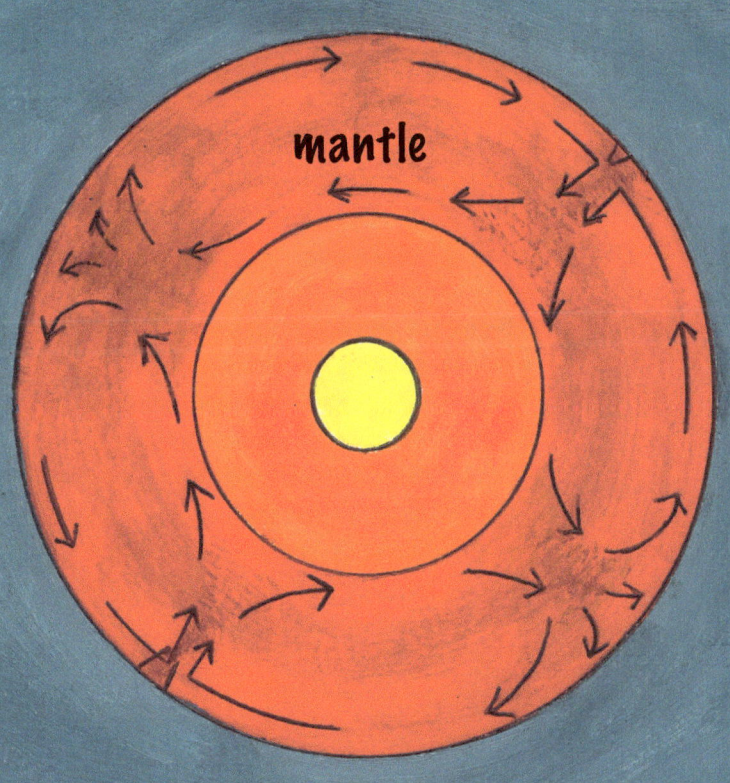

mantle

Earth

The continents that the Earth has today are fragments
of a giant supercontinent, called Pangaea,
that existed 250 million years ago.

The movement of the mantle of the Earth
has broken its crust into sections called plates.

In places where the plates are ripping apart,
new crust is formed as hot molten rock wells up.

In places where the plates are pushed together,
old crust is destroyed or simply slides past each other.

The boundaries between colliding tectonic plates are
the Earth's most active volcano and earthquake zones.

Tectonic Plates

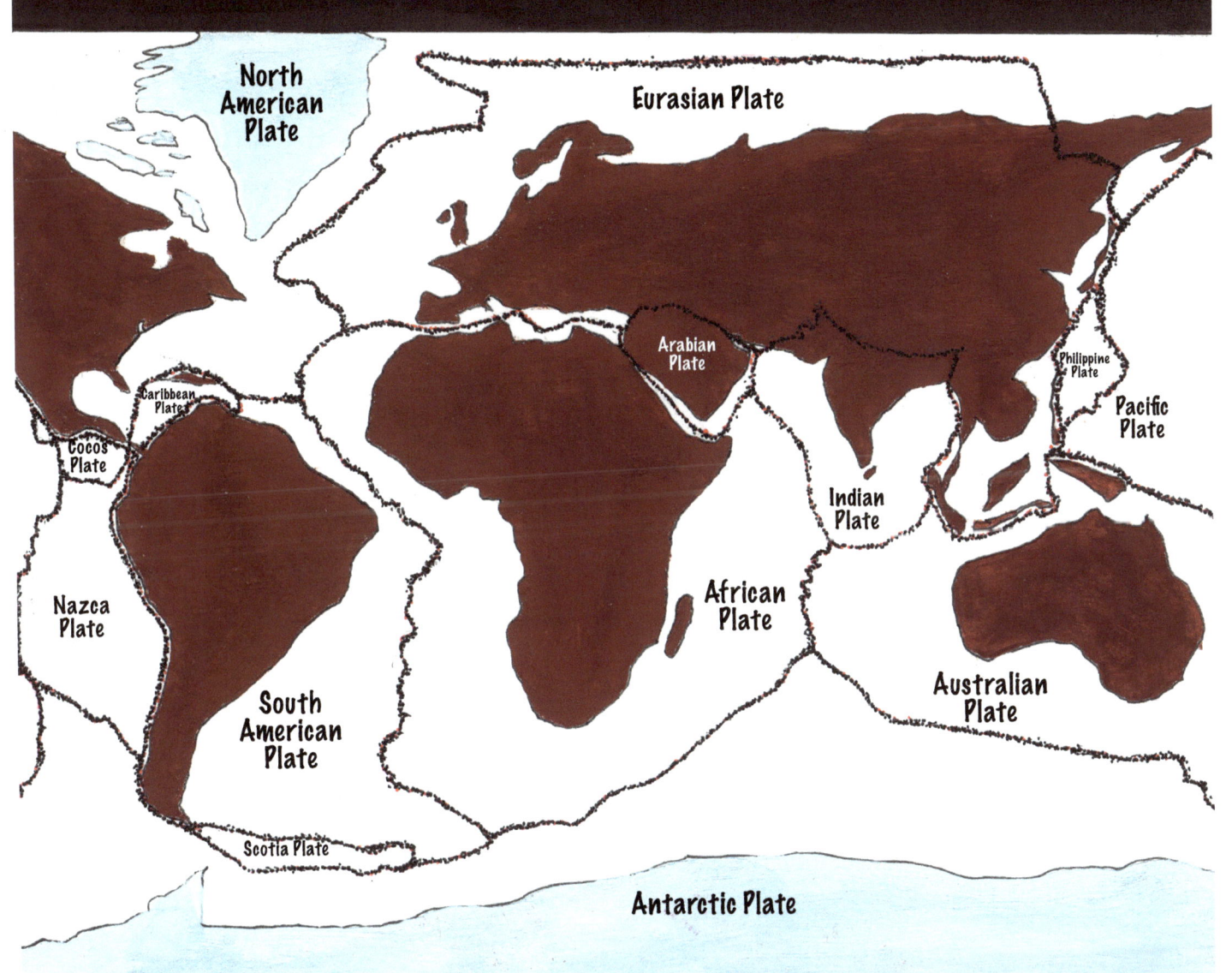

In oceans, surface water currents are generated
by wind and an uneven distribution of heat
between the equatorial regions
and the regions north and south of the tropics.

Deep currents are producd by sinking of colder water
and upwelling of warmer water.

The global ocean circulation system transfers heat
from low to high latitudes
making the oceans responsible for
about 40% of Earth's global heat transfer.

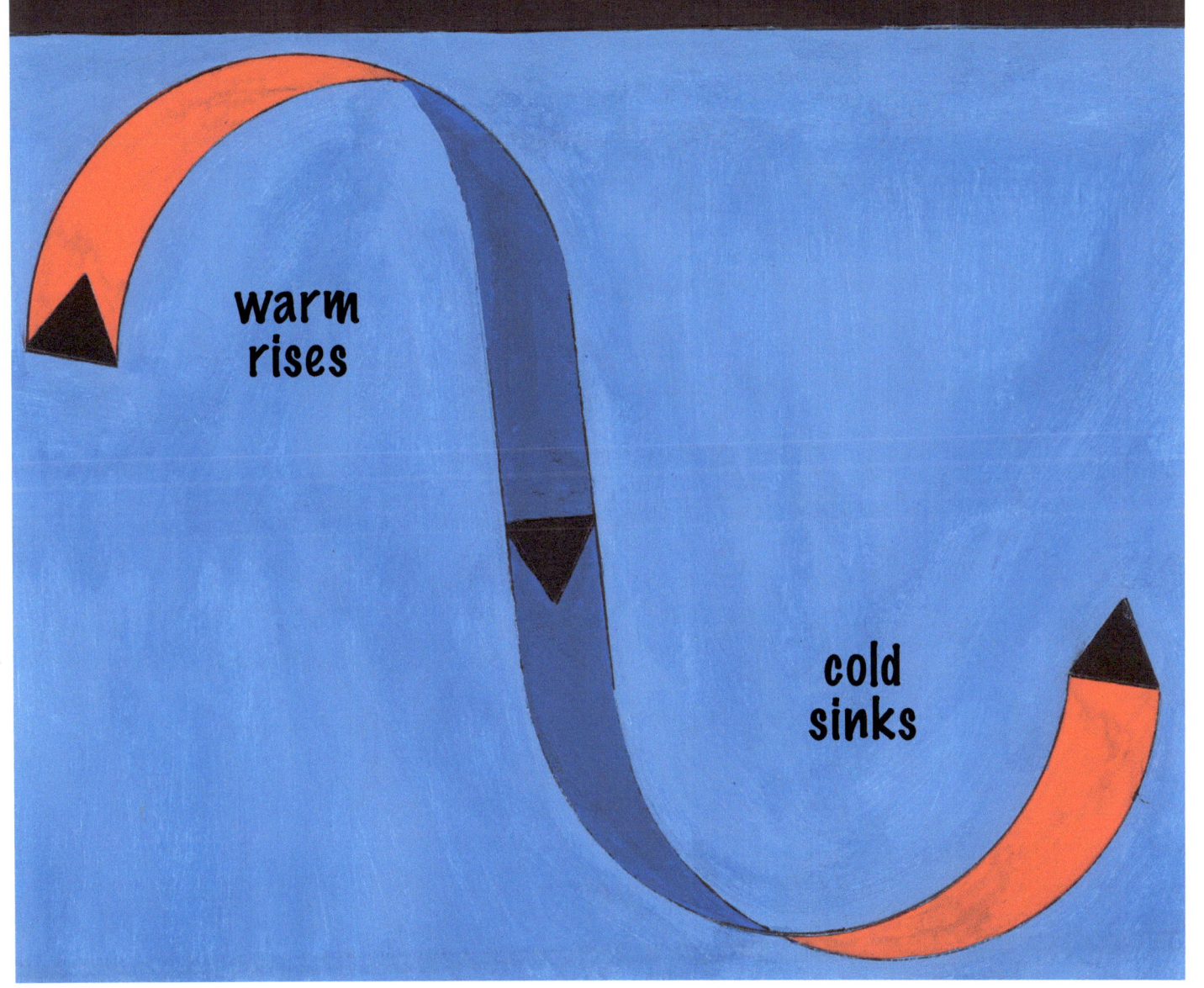

In Earth's atmosphere, certain gases such as
carbon dioxide, methane, and water vapor
act like the glass of a greenhouse.
They let sunshine through to heat the surface
but stop the heat from escaping back into space.

This keeps us warm.
Without it, Earth would be too cold for life to exist.

As more greenhouse gasses accumulate in the atmosphere,
they are increasing the greenhouse effect
and in turn global temperatures are rising.

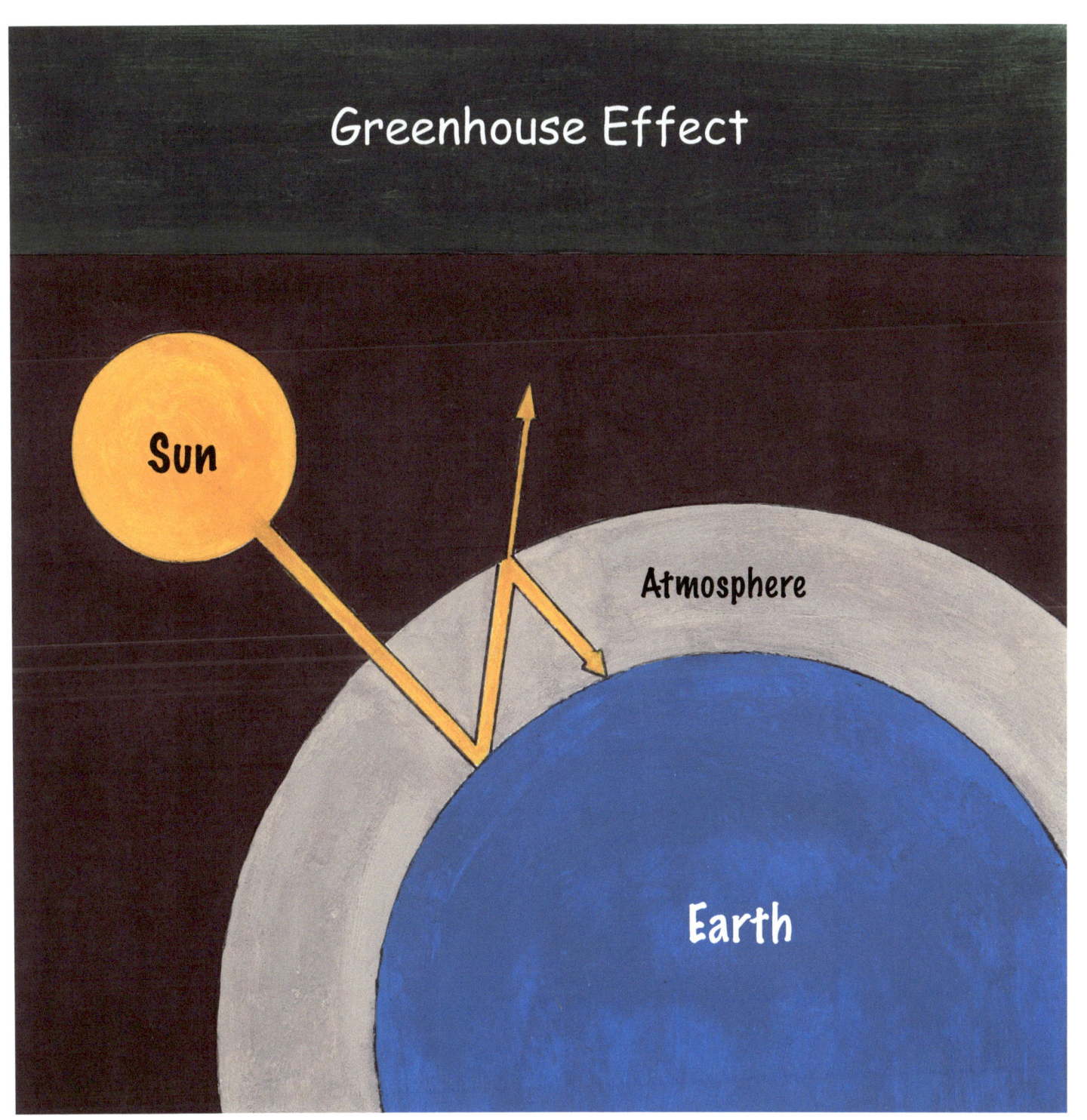

What is the Zeroth Law of Thermodynamics?

Systems are said to be in equilibrium
if the small, random exchanges between them
do not lead to a net change in energy.

A hot body in contact with a cold body gives up heat
and soon the two become the same temperature
or are in equilibrium.

Equlibrium

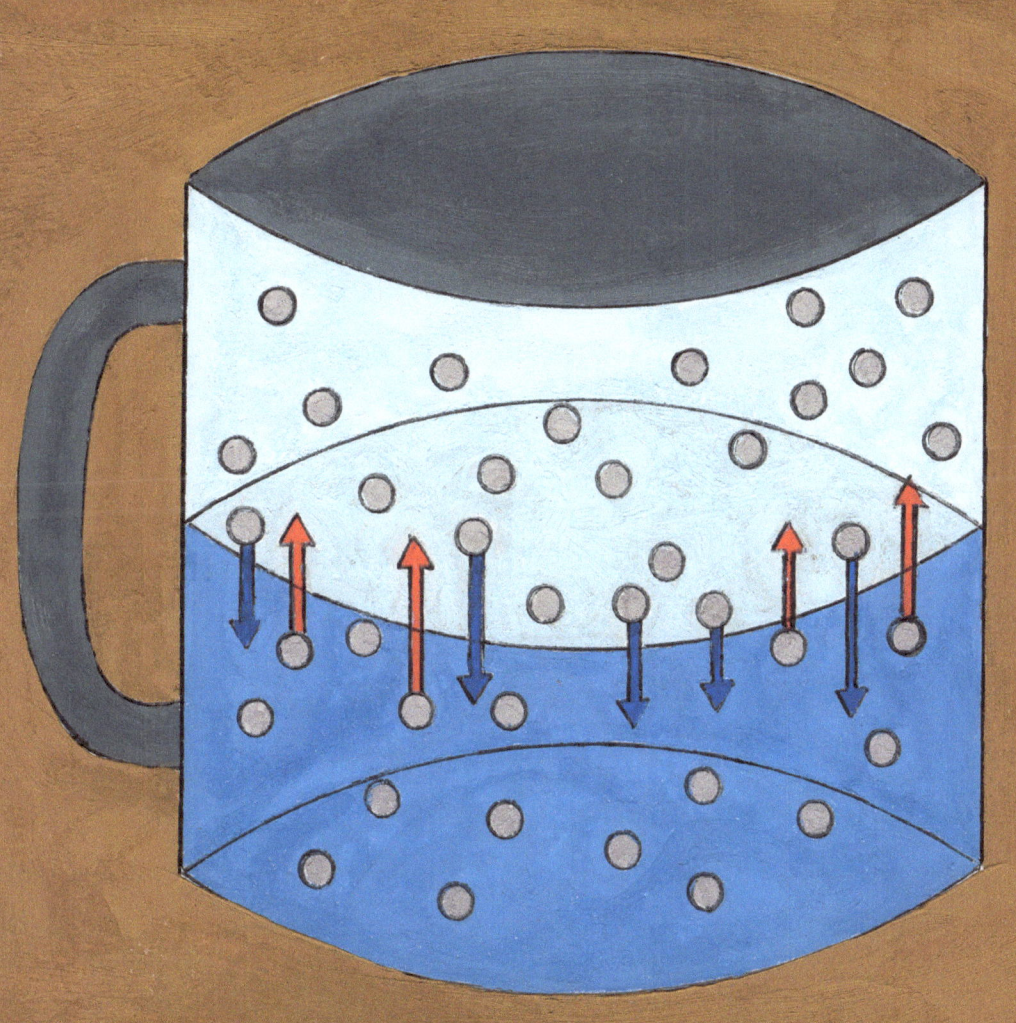

What is the 1st Law of Thermodynamics?

Energy is conserved.

The total amount of energy in the Universe is constant.

Energy can be changed from one form to another;
however, energy cannot be created or destroyed.

Interactions among fundamental particles are reversible;
one particle approaches another and then they recede
from one another.

In frictionless space, like outer space, a pendulum
is reversible and swings back and forth continuously.

Pendulum

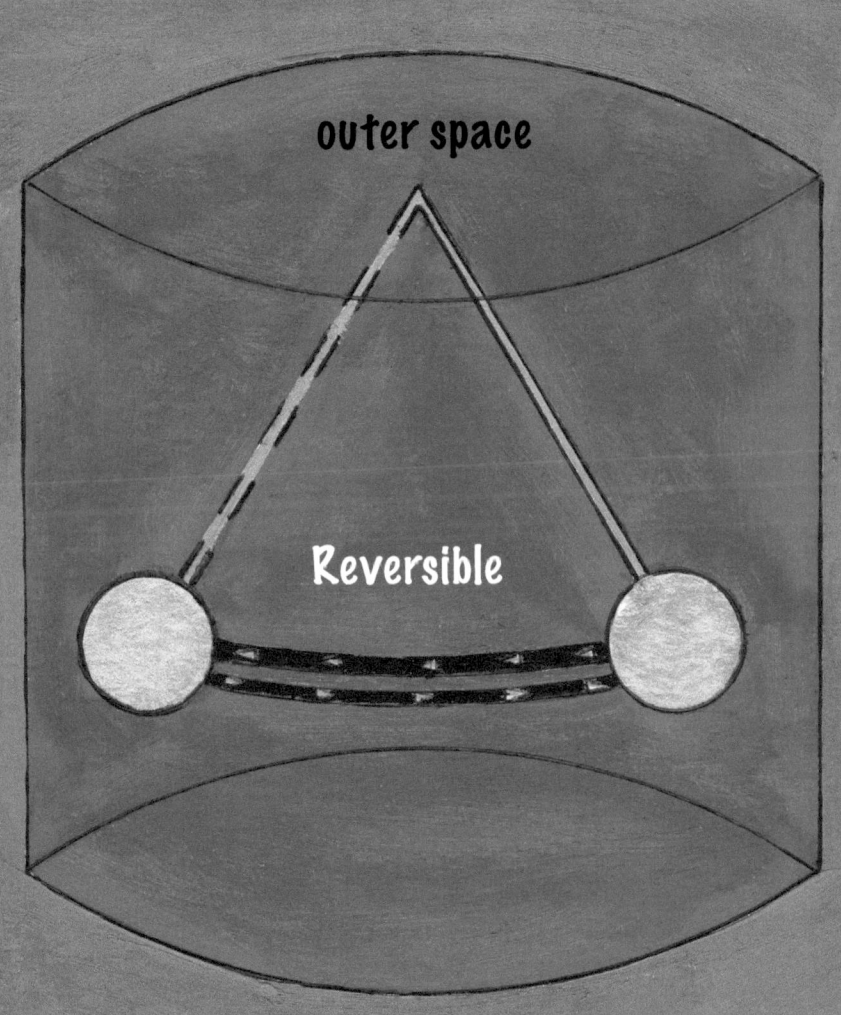

outer space

Reversible

What is the 2nd Law of Thermodynamics?

The Law defines the direction thermodynamically
a system can evolve.

Thermodynamic processes proceed in one direction;
all are irreversible.

Every time your French Fries get cold,
you experience an irreversible process.
You do not experience cold French Fries sitting on your plate
getting hot on their own.

The flow of heat from a warm body into a cool one
is irreversible.

The forward march of time is measured by the flow of heat.
Warmth disperses and time passes.

French Fries

What is Maxwell's Demon?

In order to experiment with the thermodynamics of particles in a system, James Clerk Maxwell invented an imaginary creature, a very quick and agile being, a "Demon," who could see and manipulate individual molecules.

Maxwell imagined a box that is divided into two halves by a partition with a door. The door is controlled by his Demon.
He opens the door to let the fast air molecules
go from the right side to the left and
closes it when slow ones approach it from the right.
Likewise, he opens the door for slow molecules
moving from left to right and
closes it when fast molecules approach from the left.
After some time, the box will be well organized, with all the fast molecules on the left
and all the slow ones on the right.

Maxwell's Demon showed that work must be done
to increase order and decrease Entropy.

Maxwell's Demon

Thermodynamics defines the existence of a quantity called Entropy.

Entropy is the measure of irreversibility within a system.

Thermodynamics characterizes how systems change, evolve, or proceed.

The internal Energy of a system can be changed in only two ways: heating or cooling and work done on or by it.

We know how to calculate and measure the Entropy in a system.
However, it is not local.
You cannot point to it and say "There is Entropy."

Entropy is difficult to visualize.
We know it conceptually.

Where is Entropy?

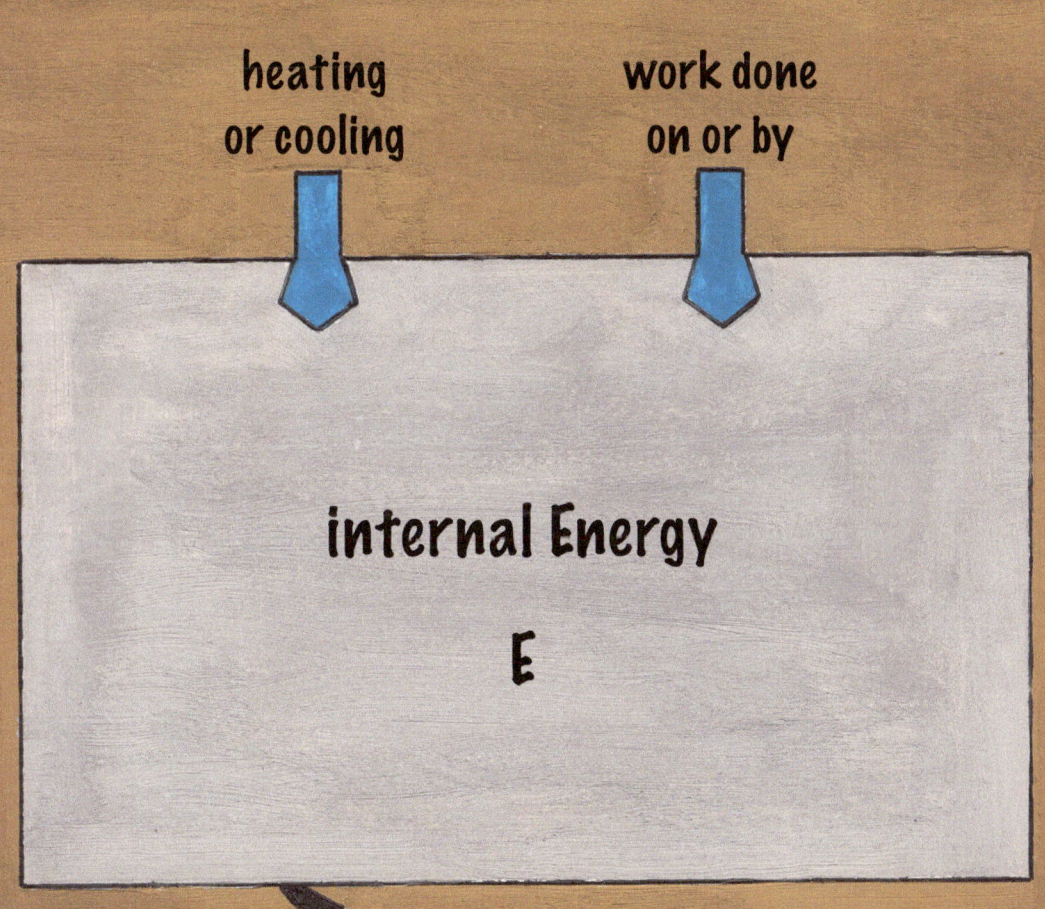

Someone or something has to do work to turn disorder into order.

Entropy will never decrease on its own unless an outside agent works to decrease it.

Entropy provides a comparison between two states.

Work

Disorder

Order

Order
Low Entropy

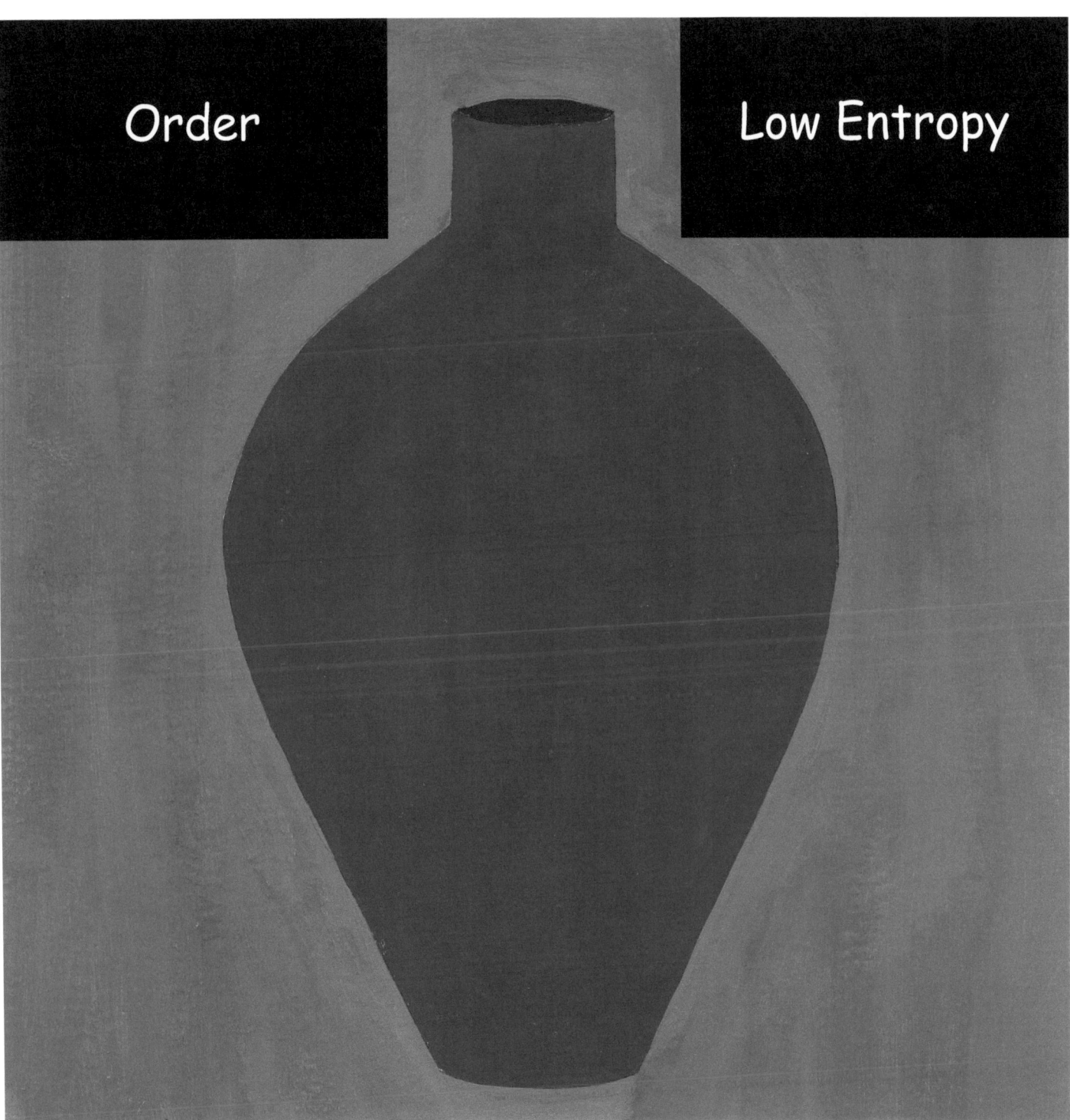

Time

Energy and entropy play different roles
within the science of Thermodynamics.

The energy of an isolated system remains constant, while
the entropy of an isolated system cannot decrease.

Entropy always increases until it reaches a maximum value.

Eventually, the different subdivisions of Thermodynamic
systems become groups of atoms and molecules.

The future is defined as the direction of time
in which entropy increases.

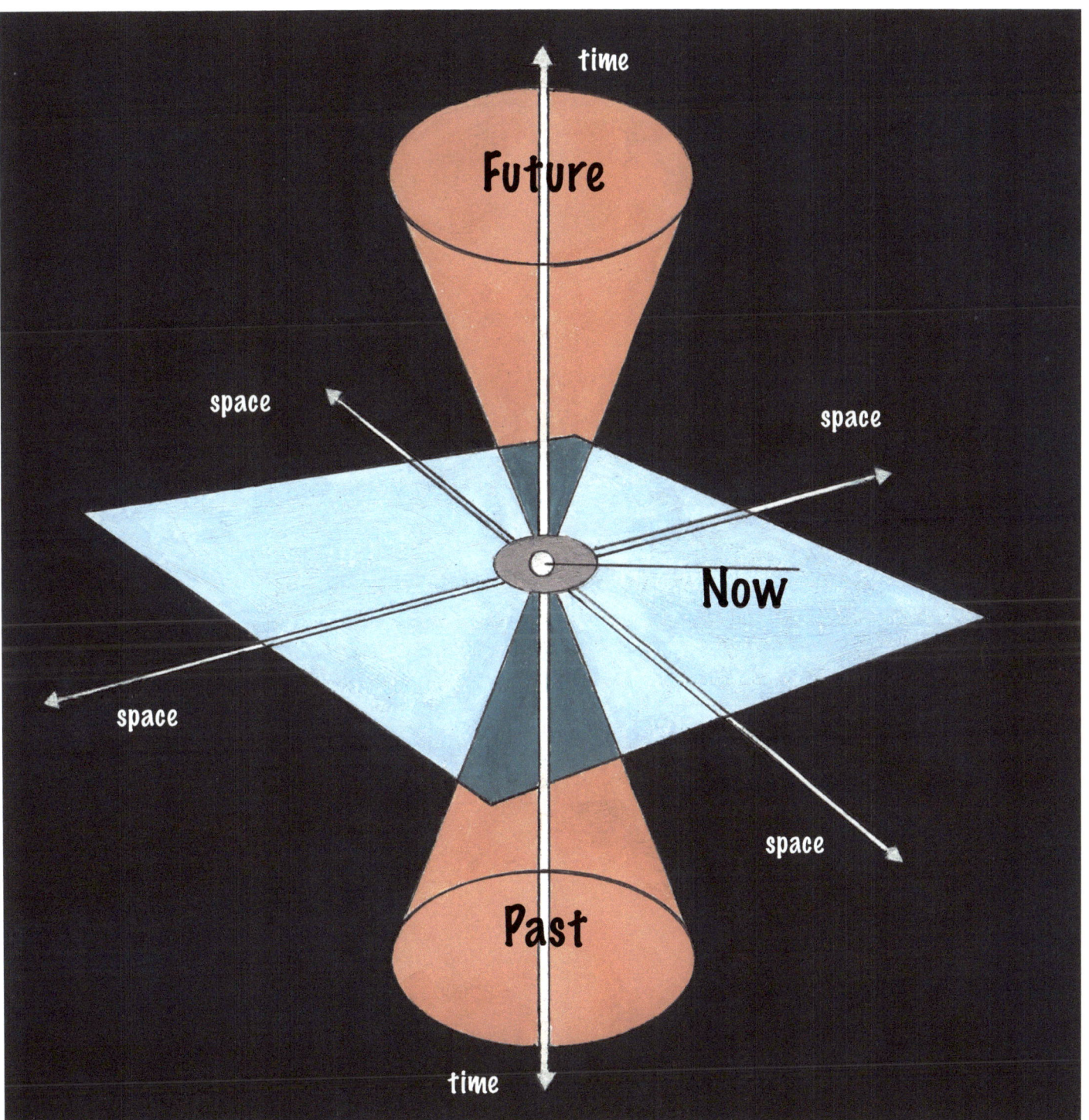

Humpty Dumpty sat on a wall.

Humpty Dumpty had a great fall.

All the king's horses
and all the king's men,

couldn't put Humpty together again.

Nursery Rhyme

Possibility

As the constraints that inform a living organism are removed,
like a tree in a wildfire,
the entropy of the organism increases.

As the tree dies, new seedlings may be coming.

Even in the death of a tree,
new possibilities sprout.

What is the 3rd Law of Thermodynamics?

The entropy of every thermodynamic system approaches zero as its thermodynamic temperature approaches absolute zero.

Absolute zero, at which all activity would stop
if it were possible to achieve,
is -459.67 degrees F.

The 3rd law is a statistical law of nature regarding entropy and the impossibility of reaching absolute zero temperature.

Absolute Zero

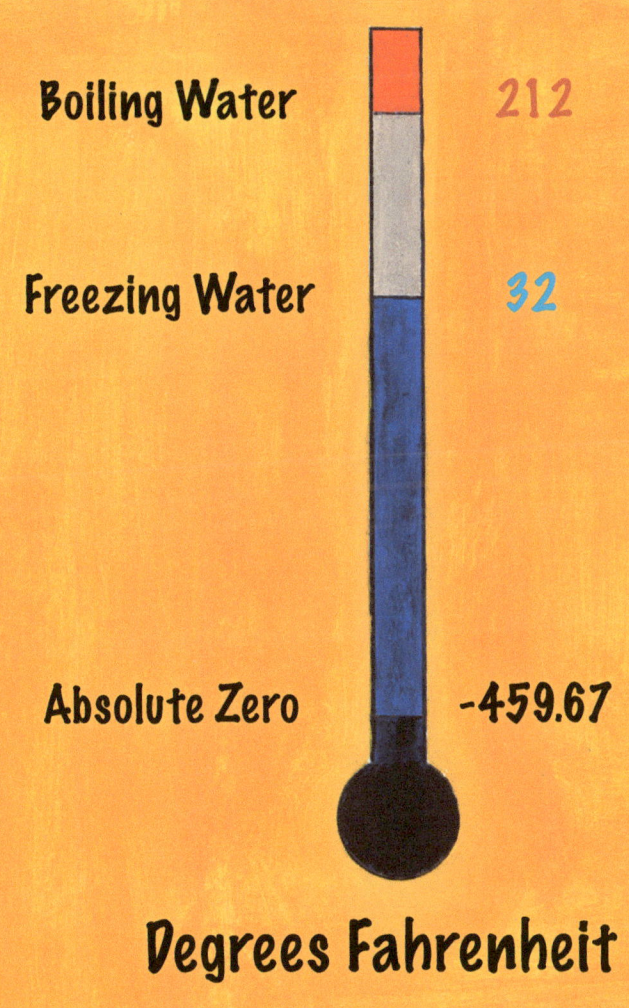

Boiling Water 212

Freezing Water 32

Absolute Zero -459.67

Degrees Fahrenheit

Science shows you can measure the energy and entropy of a system by carefully controlling and measuring the energy changed to or from a system by heating and cooling or by having work done on or by it.

The Zeroth Law requires that two boxes each in thermal equilibrium with a third are in thermal equilibrium with each other.

The 1st Law states the energy of an isolated system can never change.

The 2nd Law states that the entropy of an isolated system can never decrease.

The 3rd Law states the entropy of every thermodynamic temperature approaches absolute zero.

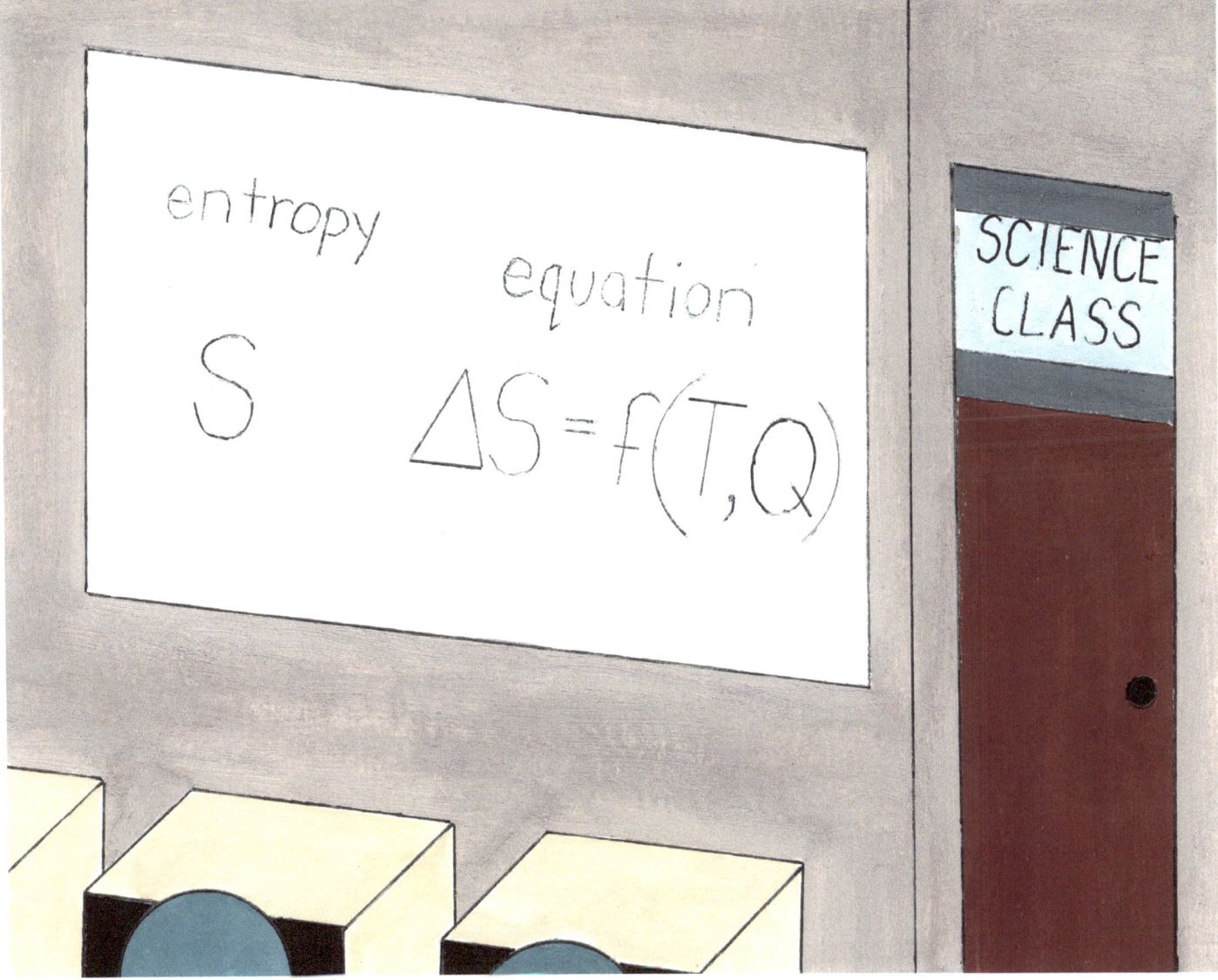

Quantifying Entropy

Entropy
S

Greek roots for the word:

"en" meaning "in"
"tropy" meaning "turn"

in turn

CHANGE

The Equation

$$\triangle S = f(T, Q)$$

A heat reservoir with temperature T is heated
by absorbing energy Q.
If a heat reservoir increments its entropy
(additively distributes over its parts)
that increment $\triangle S$
can be a function of only two quantities:
the temperature T
and the heat Q, where the sign Q
indicates energy is absorbed through heating.

HEAT RULES

Complexity

The world you live in is complex.
There are a great many independent agents, "Demons,"
interacting with each other in a great many ways.

These rich interactions all around you allow the Universe
as a whole to undergo spontaneous self-organization:
flying birds adapt the actions of their neighbors
organizing themselves into a flock, and
the human brain constantly organizes and reorganizes
its billions of neural connections
so you can learn from experiences.

The Future

Will you be the curious student
that attends Science Class
and brings clarifying or
new understanding into the world?
Even a new discovery!

Your future is full of possibility!

Original Artworks

by

Gayle Weyher

colored pencil on Bristol paper
11 inches X 11 inches

gouache on Bristol Paper
7 inches x 7 inches

Visual Learning Books

look close   look deep

visual learning

Gayle Weyher

Master of Fine Arts
Fulbright Scholar
NEA recipient

www.ingramcontent.com/pod-product-compliance
Lightning Source LLC
Chambersburg PA
CBHW040306010626
45792CB00025B/1058